Axel D. Nelke

Mass and elementary particles
Forces in the 4-dimensional space-time

Axel D. Nelke

Mass and elementary particles

Forces in the 4-dimensional space-time

Bibliographical Information of the Deutsche Nationalbibliothek:
This publication is listed in the Deutsche Nationalbibliographie of the Deutsche Nationalbibliothek; detailed bibliographical information can be accessed under http://dnb.dnb.de

© 2023 Axel D. Nelke

Printing, Production, Layout and Cover Design:
BoD – Books on Demand, Norderstedt
ISBN: 978-3-7568-3103-6

"Imagination is more important than knowledge. For knowledge is limited to all we now know and understand, while imagination embraces the entire world, and all there ever will be to know and understand."

Albert Einstein

TABLE OF CONTENTS

PREFACE

The ideas behind this book began to form in the 1970s. Quarks were yet unknown. Only the three leptons electron, muon and tauon with mostly the same properties were observed. The relations were largely unknown. There was no explanation for the mass of particles and the connection with gravity. A quantum mechanical description in terms of quantum gravity was not fully satisfactory. Nor did the simultaneous successful further development of a new physics using the theory of relativity offer a solution through quantum mechanics.

This gave rise to the idea that the common basis of the leptons could be only one particle in different states. The existing three spatial dimensions, and time as the fourth dimension, are treated equally in physics. This was followed by the idea of a new particle, the master particle, which can oscillate within narrow limits on the time axis. According to this assumption, it possesses, for example, the properties of leptons and occurs simultaneously in different energy states in an oscillation on the time axis. This could explain the different masses of leptons with the same properties and enable a quantum mechanical description of the mass of an elementary particle for the first time.

The name of this particle, the master particle, comes from the line, "Three quarks for Muster Mark!" from James Joyce's novel "Finnegans Wake", which already provided the name "quarks".

In this book, the idea of approaching the calculation of the mass of elementary particles is developed and described. Complex mathematical representations and calculations are deliberately avoided in order to make the basic ideas more understandable. To ensure a tangible representation of the foundations of theoretical physics, this format has also been favored in terms of the description.

With great respect for the achievements in theoretical physics in the last 100 years, I would like to present my idea in this work.

I would like to thank Mr. Wolfgang Sass for the encouragement to write the book and for reading the first manuscripts. My very special thanks go to Ms. Jodie Murphy and Mr. Danny Kroeger for translating the book into English.

Axel D. Nelke

INTRODUCTION

For hundreds of years, humanity has been looking to the natural sciences to explain the universe. In the search for answers, we have studied the infinitesimal, such as particles, as well as the macroscopically large, such as the universe itself. Quantum physics, in combination with the theory of relativity, tries to explain the behavior of the smallest particles.

This book is about subatomic particles, the so-called elementary particles, that are the building blocks of all matter. Thanks to countless efforts in understanding the subatomic world, scientists can deduce the properties of said particles. The objects all around us are made of atoms and molecules. In the beginning, the relationship between atoms and molecules was a mystery, just as the elementary particles are today. In recent years, however, our understanding of even the most miniscule is beginning to deepen. We are beginning to understand what lies beyond the atom and what the universe is ultimately made of.

There are different types of elementary particles. These cannot only be distinguished by their inner properties but also by the different forces they are reacting to and they themselves exert on other elementary particles.

Nevertheless, there are yet many unanswered questions:
- Why are there so many different elementary particles?
- Why do the particles have the exact masses they have instead of different ones?
- Why do they interact with each other the way they do and not otherwise?

The following new idea is a bit strange. It came to me when looking at different phenomena that are at the basis of quantum mechanics, but from a different perspective.

This book is exploring the cause of mass in elementary particles and antiparticles as well as the relationship between them. A new Standard Model is proposed. A Standard Model [1, p. 499—511] describes the summary of all properties of known elementary particles as well as the interacting forces between them. There are four different kinds of interactions, which are known as different forces. There are strong, weak and electromagnetic forces. The last known force is the weakest: gravitation.

On their own or in combination those four kinds of forces explain all interactions known to physics. Even the interactions between the subatomic particles and matter itself can be described using those fundamental forces. The same holds true on Earth, in suns, in the universe, everywhere. There are no other forces needed to explain the universe in its entirety. These forces are caused by the exchange of special subatomic particles, the bosons.

There is another category of elementary particles, the so-called fermions, which these forces are affecting. Together, fermions and bosons are the sum of all elementary particles. They all have different masses or sometimes no mass at all. An explanation of the physical cause of mass of an elementary particle is only possible with limitations and is not mentioned in the aforementioned Standard Model. Furthermore, gravitation explained by the gravitons as bosons is non-existent in the Standard Model.

The same goes for dark energy and dark matter. At the end of this chapter, these two terms will be discussed and explained. The newly proposed Standard Model takes all these points into consideration.

Especially important are the above-mentioned antiparticles. There are always two of each kind of elementary particle: the normal one as well as the antiparticle. Due to the complete symmetry, antiparticles can combine to antimatter just as normal particles can combine to matter. When a particle and an antiparticle of the same type come into contact, both are annihilated and dispersed as photons. Vice versa, a photon can be converted into a particle and an antiparticle; physicists call this process pair production.

The next chapter will explain all fermions and bosons of the Standard Model in greater detail.

Contextually, it is important to look at the effects of gravitation on the space-time structure according to the laws of general relativity [2, p. 177—224]. In addition, it is also important to observe the effects that velocity has on a particle, which is explained in the theory of special relativity [3, p. 3—162], [2, p. 98—176]. Quantum mechanics [4], [5], [6] will also play a big part.

For a better understanding of the concepts of this book, some important points will be extracted from those sources.

The sum of energy will always stay the same in the entire universe, and any given physical experiment or process will result in an equilibrium of energy. This is known as the conservation of energy, the first law of thermodynamics, a fundamental premise of physics. Independent of the three directions of space, energy can exist in different states. However, energy cannot be transported to the past or the future, as this would contradict the first law of thermodynamics (energy conservation law). There is potential energy, which is stored energy that can be converted into other forms of energy when the object's position or state changes, for example when an object near Earth falls towards it. The potential energy is then transformed into kinetic energy. Energy is not only made up of kinetic and potential energy but can oftentimes exist as a combination of them. An example illustrating the combination of kinetic and potential energy is seen when observing the vibrations of a guitar string. Another example is the oscillation of an electron around an atom. Potential and kinetic energy alternate.

The mass of a particle is another example of energy, known as rest energy. From Einstein's theory of relativity, it is widely known that mass is energy and vice versa [2, p. 149—151].

The energy of electrons in atoms is bound to certain energy levels [7, pp. 110-131]. An energy level is the discrete energy, which is part of so-called eigenstates in a quantum mechanical system such as atoms. These systems can permanently only be in one such eigenstate but not between those energy levels. The lowest state is known as ground state, whereas anything above that state is known as the excited state.

Inside an atom, it can thus be deduced that electrons can only take certain eigenstates. Consequently, there are only discrete spectra of energies of electrons inside of an atom and not continuous spectra. An electron can change its eigenstate to a lower state. This process releases energy. Conversely, when an electron changes to a higher state, energy can be stored.

Only through quantum mechanics can this behavior be understood [5], [6]. It can be explained in mathematical formulas of quantum mechanics in which the eigenstates are described as a stationary wave function. These are quantum states in which the wavelength, adjusted to the physical conditions of the environment, will always fit into an integer value. Anything in between is not possible.

The quantum mechanical phenomenon of the wave properties of a particle is explained by the interference with other particles and with itself, as known from the double slit experiment [4, p. 107—113]. The wave character of particles is revealed by the superposition of their waves. This experiment demonstrates the duality of a particle and shows that a particle is both a wave and a particle, known as the wave-particle duality. Classical waves propagate in a given space. Another wave, through interference, will add or subtract from that wave, whereas this interference can happen at multiple different points in space simultaneously. A classical particle can only be in one point of space in a given time. This seems to be contradictory. Still, through the double slit experiment, it is shown that different quantum objects show both properties, such that they can also be described as a matter wave. If a particle, however, interacts with or is observed by another particle, the wave function collapses and the particles wave properties disappear.

Now something completely different: If a particle and its antiparticle come together, they are both destroyed, which releases energy. The amount of this energy is calculated by using the known formula first introduced by Einstein: $E = m \cdot c^2$ [2, p. 149—151]. Conversely, a particle and its antiparticle can spontaneously come into existence in a vacuum. This is called pair production [8, p. 20—21]. It is constantly happening in vacuum and is known as quite normal within particle physics. Through this observation, it can be deduced that the energy of a particle and its antiparticle can be described as a wave, which can destructively interfere and release energy in this process. The only remaining question is: Which part is oscillating, and which part carries that oscillation? It is clear that there must be an almost massless particle, which is necessary as a basis for an oscillation. It is also important because it must carry the later properties of the elementary particle. So, it can be assumed that there is an even smaller particle as basis.

In the whole universe, there is a continuous stream of neutrinos that are created by suns, penetrate everything and barely interact with other particles. That's the reason for the name. On Earth, there are roughly 100 neutrinos present within the volume of a matchbox. They penetrate everything without leaving any trace. These particles, neutrinos and antineutrinos, would be suitable to carry the oscillation.

Now we only need some phenomena of quantum mechanics. A small amount of energy can actually be borrowed for a short period of time, only to then disappear again. This phenomenon has its origin in the uncertainty relation between energy and time, a relation that comes from quantum mechanics and is described by the formula: $\Delta t \cdot \Delta E \geq \hbar$. Another uncertainty relation exists between the location of a particle and its velocity, described as: $\Delta x \cdot \Delta p \geq \hbar$. These relations are known as uncertainty relations [8, p. 72—76] and are an expression of a dualism between wave and particle in a four-dimensional reality. As a result of the uncertainty principles, a particle can never be fully described given its velocity and location or energy and the discrete time of that energy. If the location is certain, the velocity is uncertain, or if the concrete time of a given location is known, the energy state is uncertain.

An amount of energy can be borrowed for a short time without invalidating the law of conservation of energy. This means that it is possible for a particle to borrow so much energy over a short period of time that it can overcome an otherwise insurmountable obstacle. The period is extremely short and, depending on the general conditions, can sometimes be less than a billionth of a second. This is also observed and would normally be inconceivable in classical physics. The aforementioned pair production (particle and antiparticle) is also made possible by a so-called fluctuation of energy via the uncertainty principle (between time and energy). At any time, so much energy can be borrowed that a virtual particle pair is created, which, however, must disappear again after a certain period of time.

Every force that two particles exert on each other is caused by the exchange of a special particle. In electromagnetic interactions, this force is triggered and transmitted by photons. Photons can be detected, and their properties can also be calculated mathematically. As already mentioned, this type of particle for the transmission of force is called a boson. The same applies to the gravitational force. The known mathematical and physical properties of gravity are attributed to the exchange of gravitons, only these have not yet been observed, although they should be present in very large quantities. It is, however, undisputed that they exist. This is an important starting point for the hypothesis presented later in this work.

The speed of light in a vacuum is constant and unchanging, regardless of the physical conditions. It is also a speed that cannot be exceeded, and it fundamentally describes a fixed ratio between the units on the time axis and the space axes. Photons moving through space at the speed of light have the maximum possible speed and move on the boundary between space and time. For further explanations, some points from the basics of the theory of relativity are presented summarized.

We live in a dynamic space-time continuum, which is explained through the effects of the speed (special theory of relativity) of particles as well as the gravitational influences (general theory of relativity) within the four-dimensional space [2] [3]. The result of that has basic consequences. An acceleration that influences a particle, such as gravitation, will result in a slowdown of the proper time of that particle. The same goes for the speed of that particle, which has the same effect. At the same time though, the axes change. In contrast to the time axis, which apparently expands, the space axes contract.

The consequences can be directly observed in a moving electrical charge. Electromagnetism is the name for an interplay between electric and magnetic forces, which depend on the charge and the velocity of a particle. A resting charged particle only has an electrical field but no magnetic field. If this particle is now set into motion, it begins to develop a magnetic field in addition to the electrical field. The direction, size and orientation of said fields are explained through the Maxwell equations [9, p. 208—221]. This behavior can be explained due to the changing time axis, as explained by the special theory of relativity. This leads to the electric charge taking on a new property, which manifests itself as a magnetic field. We will revisit this idea again later and explain it in more detail.

These correlations are explained mathematically by the special and general theory of relativity. Those mathematical formulas are only important as a side note in this book. The following paragraph will be more important than the formulas.

Exceeding the speed of light is unattainable in our "real space" and a prerequisite in the theory of relativity. Something moving faster than the speed of light is not observable and can thus not interact with us. It is, so to say, not feasible. However, mathematically, there is no objection to a particle surpassing the speed of light. The known physical laws and even the conclusions of the theory of relativity are mathematically transferrable to superluminal velocity. It is thus possible that there is a parallel reality without any reference to what we perceive as the real world. This idea is fundamental to creating the working hypothesis of this book.

The following segment concentrates on explaining the terms that are often used in this book and thus need a detailed understanding.

The spin [8, pp. 93-97] of a particle is an unchangeable property, which means an intrinsic kind of angular momentum. It is described with the Planckian quanta of action, $\hbar$.

There are four interactions or basic forces of physics, triggered by the exchange of bosons:

- Gravitation comes into effect between two bodies with mass and also affects energy (gravitons)
- The weak force is the basis for the force between certain elementary particles and enables some particles to transform into other particles (W particles)
- The electromagnetic force is known for the forces between two charges as well as from magnetism (photons)
- The strong force, or strong nuclear force, is responsible for the stability inside of the nucleus (gluons)

The parity conservation [8, p. 214—221] describes how two physical systems, of which one is the mirror image of the other, must behave identically. The only exception to that rule lies in the weak forces in which some mirror images do not exist. The violation of parity is exclusively and only known for the weak interaction.

The lepton number conservation law states that before and after any physical process of particle interaction, there must always be the same number of leptons and antileptons. The explanation for these interactions will be discussed in the next chapter.

The same holds true for the number of quarks in which there will always be the same number of quarks and antiquarks before and after the process of particle interaction.

Dark matter is a presumed form of energy or matter that is not directly visible but interacts through the gravitational force it exerts. Its existence is suspected and is the only way to explain the movement of the observable matter in the universe. The size and velocity in which stars rotate around the center of their respective galaxies can only be described with the existence of dark matter. In the outer space of galaxies, the velocity is much higher than can be described by the gravitation of observable stars, gases and dust clouds. To our current understanding, only a small part of matter can be observed and is thus documented in the Standard Model as elementary particles [1, p. 501—504], [10].

Dark energy is a hypothetic form of energy that describes the observed accelerating expansion of the universe.

PARTICLES AND FERMIONS

Figure 1 summarizes the elementary particles in a tabular fashion. In this table, the fermions and bosons, the two basic forms of elementary particles, are separated into groups.

The top two rows (groups) of fermions represent the quarks, and the bottom two rows (groups) represent the leptons. We will look at the bosons later and put them into relation to the fermions.

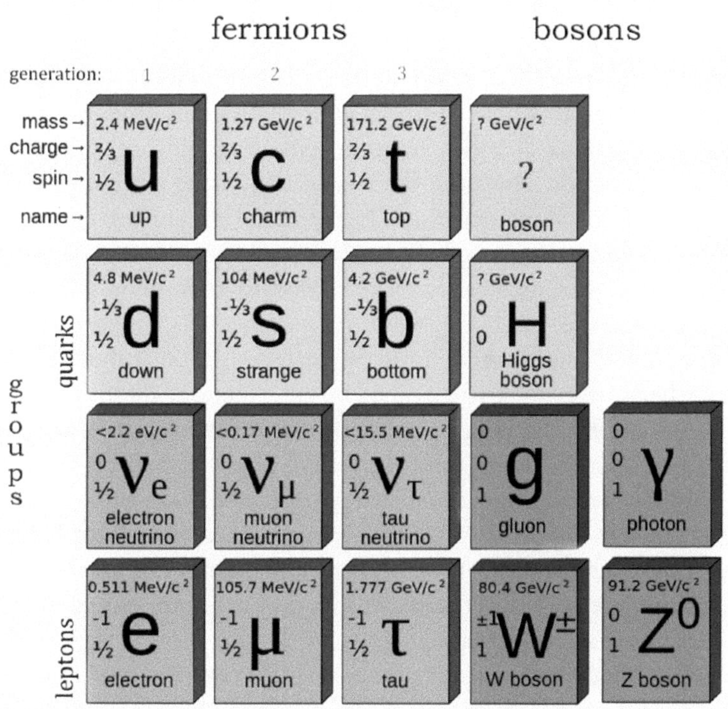

Figure 1: The Standard Model of particle physics

The name fermion comes from the physicist Enrico Fermi. Three quarks from the two top rows with the three different generations make up many other particles, such as neutrons (one up and two down quarks) and protons (two up and one down quarks), which are both part of the nucleus. The u in the top row stands for up quark, followed by the charm and top quark in the same row. The mass is deductible from the energy that's noted on the top of each particle's box and can be calculated with $E = m \cdot c^2$. The different mass of fermions is the main difference between the different generations with otherwise the same properties. The second row depicts the down, strange and bottom quarks.

The third and fourth rows likewise make up the three generations of leptons. The three different generations of leptons are called electron (e), muon (μ) and tauon (τ). The third row contains the neutrinos (ν). As we can distinguish three different kinds of neutrinos, there have to be other, yet unknown, distinguishing properties besides the mass. However, there is an interesting additional piece of information about neutrinos. Neutrinos of the three generations, given a known neutrino oscillation, can transform into each other, whereby we can deduce that a reduction to a common particle is not jeopardized.

In the two rightmost columns we can see particles that act as a force between two particles called bosons (photon, gluon, Z and W and on its own the Higgs particle or Higgs boson (H)). These particles are named bosons after the Indian physicist S. Bose. Photons and gluons are shown without energy (eV), as they are without mass. It is possible that gluons have the same value as neutrinos. Permanent transformations between gluons and quarks can be observed, which may lead to the conclusion that they are more comparable to neutrinos than real bosons.

Something worth noting in the table is that there is another boson found in the top right corner, the existence of which is assumed due to this book. It would be a new boson that is yet unknown. We will get into more detail later.

In summary, the table shows a similar grouping as the periodic table does in chemistry in which particles with the same properties are grouped together. Fermions with quarks and leptons with the same spin and charge only differ from generation to generation in growing mass. Elementary particles with equivalent properties in the three generations are called groups in this table.

The growing mass in the different generations of fermions is similar to the different energy levels of a quantum mechanical model of oscillation, e.g., different energy levels of electrons in a hydrogen atom [11], [7, p. 110—131]. Atoms are made up of a heavy, positively charged atomic core and negatively charged electrons that orbit the core. The energy levels of different orbits are called orbitals. They are described as discrete quantum mechanical energy levels and are characterized by quantum numbers. An electron can change from one orbital to an orbital with a lower energy level. This process releases a photon, which carries a frequency that is equivalent to the energy difference between the two orbitals.

Put simply, this book assumes that these particles are made up of smaller particles. As such, there must be smaller particles as a basis, which carry the same properties (spin and charge) as the groups (e.g., electron, muon and tauon). This even smaller particle oscillates in different energy levels and makes up the three particles of that group.

Especially in recent years, various models have been explored to develop new theories in the realm of elementary particles, postulating that they are composed of even smaller fundamental constituents. This book will assume that all three fermions of a given group can be described given one yet-to-define particle that is simply in different oscillation states. For the sake of this book, we will call these new particles "master particles". The master particle would have all the properties of a group of the three different generations. This way, the properties spin and charge will be the same for these three fermions. Only the mass of a given fermion would be different due to the different oscillation states of the new particle with its given energy. These master particles would exist separately for each group. As such, there would be three different master particles: one for all of the leptons (electron, muon and tauon and their neutrinos) and one for each of the two groups of quarks. It would even be possible that the two groups of quarks could be made up of only one master particle. An indication of that can be found when looking at the conservation number of quarks and antiquarks. These are constant in any particle interaction for all quarks, independent of their respective group.

The kinetic energy in the oscillation states of the master particle would be equivalent to the energy level and this to the mass of the respective fermion.

The fundamental concept of this book is the assumption that there exists exactly one master particle for each fermion group, which can be in three stable oscillation states. This would explain why the properties' spin and charge are the same for the three fermion generations in each group. How big this master particle would be, and which mass and properties it would have, is interesting but not yet to be described. The energy levels would be stable if they would sustain a self-oscillation as can be found in an atom or a stationary wave. There is even an indication that this hypothesis holds true. As is known by many particle interactions, the number of leptons and antileptons (electron, muon, tauon and their neutrinos) is always constant in any particle decay or interaction. They can always be transformed into one another and in the end, the loss of a charge leads to an interaction with the corresponding neutrino of the respective charged lepton. This could be an indication of one single particle being a basis for all leptons. This would then be the master particle for leptons.

QUESTIONS

1. What would a quantum mechanical explanation of the mass of fermions (both leptons and quarks in three generations) look like?

2. Is there a possible explanation for the missing proof of gravitons?

3. Are there any indications of a theoretical relation between dark matter, dark energy and known or unknown particles?

4. A renormalizable quantum mechanical explanation of gravitation with gravitons is not possible as there is yet no known way to describe an interaction or an oscillation [12, p. 177—191], [13, p. 1—24]. Will taking the time axis into account help to describe a new quantum mechanical model?

5. Implementing gravitons into the Standard Model has not happened so far. Which conditions need to be met in order to do so?

WORKING HYPOTHESIS

Understanding the idea and quantum mechanical phenomena will be made possible under the hypothetical idea that these corpuscular master particles have the ability to travel back and forth on the time axis and thus travel to the past and the future multiple times. This way, all possible reaction pathways are exhausted. Only this way, particles can exist in multiple locations at once, but given their probabilities of locations, they only exist once. With this new idea of traveling on the time axis, we can get a better understanding of the uncertainty principle. All possible states exist at once and so it is possible to breach the potential barrier or to break the energy balance for pair production with a given probability by adding the time axis.

At the basis, we require a four-dimensional space, consisting of three spatial dimensions and one time dimension. This is especially important as the general and special theories of relativity were created under the prerequisite of this four-dimensional space. This leads to exact results for calculations. The Riemannian geometry is imperative as a basis for the general theory of relativity [14]. This differential geometry describes and examines the geometric properties of curved spaces with the methods of differential calculus. With it, all structures, forms and changes in a four-dimensional space can be calculated. Given this differential geometry, the change in position of a particle in a four-dimensional space can be described on all axes (three spatial axes and one time axis), especially for this hypothesis, on the time axis.

The possibility of a new, undiscovered particle that travels on the time axis is thus not ruled out. Thus, a new hypothesis can be developed. As discussed previously, this particle will be referred to as the master particle, which can move on the time axis. It might even be a neutrino, but as this cannot yet be proven and is only an assumption, the name master particle will be used. Another yet unanswered question is: which potentials or forces lead to a movement on the time axis? The gravitons might play an important role in this. In this hypothesis, they exert a force exclusively on the time axis, one that cannot be exerted on any of the three spatial axes. That could prove why they cannot be observed in our reality, as they can only interact and be found on the time axis. Yet they do exert a force, so the next question is: between which particles?

Figuratively speaking, the whole universe with its mass in the three spatial dimensions is embedded into a four-dimensional space. Each particle in the entire universe interacts with all the other particle due to the exchange of gravitons on the time axis, the fourth dimension.

However, everything exists in a thin layer of time: the present. Due to the energy conservation law, no energy or matter is forgotten in the past or carried over into the future. This means that everything—energy and mass—only exists in the thin layer of time that is the present. To illustrate this, let's remove one dimension and imagine the universe as a tennis ball. The center of the ball is the origin of the time axis. The thin wall of the ball is the present with two dimensions in this case (three dimensions in reality), which contains the universe. A mass creates interactions with gravitons and exerts an attractive force on another mass. In our example of the walls of the tennis ball (the mass of the universe), the gravitons can exert a force in both directions, inwards and outwards, into the past and into the future. If such a particle would try to leave this thin layer, it would be pulled back. Only if it is in this thin layer, it would not be subjected to the forces, as they do now exist on the spatial axes. The master particle, however, would be able to oscillate on the time axis, always going through the thin layer of the present. This oscillation would bind some energy. Now, this model is nearing the prerequisites for a quantum mechanical explanation of an oscillation with the gravitons as the interacting particles.

The hypothesis of this work assumes that the mass of a fermion of the Standard Model in our reality is a result of the energy of this oscillating master particle in a parallel reality. Parallel because it travels at superluminal speed, and it thus only exists in that segment. Now we need to see what such a construct would look like.

A change of position on one of the space axes in relation to the time axis will result in a velocity v. A change of a position on the time axis could be explained by the inverse dimensioned velocity v_{ct}. This velocity (s/m) would have the inverse of the known velocity in our physical space v (m/s). This movement on the time axis can be permitted without changing the geometry or having limitations in the mathematical prerequisites. Using the speed of light as a barrier, space-time is split into two segments or areas [3, p. 89—107]. This is a logical conclusion of the theory of relativity. As mentioned, all physical laws, including the theory of relativity, are still valid in a superluminal space. Only the velocities and scales are presented in inverse. It is postulated that the same correlations of the special theory of relativity exist in inverse. The inverse relation between the time-like segment (our world) and the space-like segment (parallel world) only refers to the parameters of velocities and not the parameters of mass or energy of a particle.

The model presented here is based on the assumption that master particles always possess superluminal speed. Such a speed would mean that the master particle could move on the time axis and the maximum speed would converge to the inverse value of the speed of light. Under these premises, it is assumed that the master particle would be able to leave the thin layer of spatial dimensions to travel into the past and future within small bounds. It would be subjected to a force in a gravitational field, which would be exerted by the mass of the universe, which pulls it back to the three-dimensional space. This would accelerate the particle in the direction of the thin layer of the present of the universe, which is also known as hyperplane. Mathematically, the word hyperplane describes a plane with more than two dimensions. In this thought experiment, the master particle would breach the three-dimensional space due to its speed. On the other side, the master particle would decelerate when leaving the plane as gravitation pulls it back. This process continues, which in turn creates an oscillation.

In addition to the oscillation, there exists an angular momentum in a plane, which is created by the time axis and one or more spatial axes. The axis of rotation is a spatial axis. In the three-dimensional space and under the influence of the angular momentum, the master particle would not oscillate through one exact point. Due to the angular momentum, the master particle would breach the walls of the present at different points within a small range around its center. This behavior can be closely linked to the angular momentum of a particle, its spin. This means the master particle oscillates on the time axis and due to the rotation around at least one of the spatial axes it encloses a small area or space. Within this small space, the master particle, through its spin, can only be observed with extremely high energy inputs. This is a consequence of the law of the conservation of angular momentum.

This means the master particle would not be able to exist below a certain radius around this center in a three-dimensional space. There is, however, no limit on bigger radii, only a decreasing probability with growing radii. In our four-dimensional reality, this master particle would be present in multiple locations at once with a certain probability, which is exactly what Heisenberg's uncertainty principle already describes. The hereby mentioned rotation is to be strictly distinguished from a rotation around the time axis. Through this rotation, the electric charge in this hypothesis is defined, which will be clarified later in the book.

With this working hypothesis, another explanation of the uncertainty principle is possible as an expression of the angular momentum of a master particle breaching the three-dimensional space. Furthermore, the movement of a master particle on the time axis would explain tunneling. The master particle, with a detour through the time axis, would breach the potential barrier or obstacle. The particle would be able to travel from one location to another through the time axis without the need to breach the potential barrier on the spatial axes.

The newly introduced oscillation on the time axis could create standing waves with different "frequencies". With that, different stable energy levels can be explained. At a later time, we have to think about how the energy that is bound to the oscillation of a master particle in the parallel world can be associated with dark matter. Accordingly, it is necessary to establish connections between the different stable oscillation levels or states and the different fermions.

The mass of this master particle, which contributes towards the mass of fermions, will not be looked at in this work.

An exact idea of how the oscillation energy of a master particle can be associated with the mass of single fermions will be discussed in a designated chapter.

ON THE ELECTRIC CHARGE
OF A FERMION

The working hypothesis can help to understand Maxwell's laws of electromagnetism in a new way. The equations describe how electric and magnetic fields and how electric charge and current are connected under certain conditions. The equations are a special system of linear partial differential equations of the first order.

Due to the oscillation of the master particle on the time axis, it is able to rotate around the time axis over the space axes. This rotation could take place in a plane of at least two spatial axes. All combinations would be possible: x and y axes, x and z axes, y and z axes and all three axes. However, even a rotation over two or three planes would be imaginable. Through this rotation and its interacting forces, electric charge could be explained as angular momentum. The size of the charge would be the full elementary charge if the rotation would take place over three spatial axes. A third of the charge, as can be observed with some quarks, could be explained through the rotation over only two spatial axes around the time axis. Additionally, a rotation over two planes would equal a two-thirds charge, as can also be seen in some other quarks. Through this thought experiment, it can be explained why an electric charge does not take any higher or even any other values.

It gets even more interesting when we look at a charged Fermion that moves with a velocity, v. Through this motion, due to the theory of relativity, the property of the time axis is changed. Geometrically, the axis is tilted compared to the time axis of a resting Fermion. This tilt would take the rotating master particle out of the plane of the present. The rotational path would lead above (the future) and below (the past) our observed plane (the present). Magnetism would be an expression of the force exerted by a charge, which is stretched on the time axis of a given distance and which is oscillating. It would be possible to explain the magnetism of a moving charge. Therefore, magnetism is the expression of an electromagnetic force that exists due to the change of the time axis. As the time axis of a moving particle is tilted, the spatial axes will be involved in the rotation, which previously would only happen on the time axis.

The rotational direction of the master particle around the time axis would equal the rotational direction of the magnetic field around the moving direction of the corresponding fermion. This would correspond very well to the phenomena of magnetism described by Maxwell. It would also explain the absence of a magnetic monopole, as the rotation is self-contained. A magnetic monopole would be a single magnetic north pole without a south pole and vice versa. There is also no actual magnetic charge. This could indicate that the working hypothesis is correct. Only a moving electric charge creates magnetic vortex fields with closed field lines. The intensity of the magnetic field depends on the velocity of the moving charge [9, p. 214]. A magnetic field is always self-contained with a magnetic north and south pole. The magnetic field is perpendicular to the direction of the charge and rotates around it.

As previously discussed, in this hypothesis, the electric charge equals the rotation around the time axis, and a tilt of the time axis due to the velocity of a charge leads the charge to leave the plane of the present. Thus, the rotation around only the time axis is transformed into a rotation around the time axis and one space axis, which include the area above and below the plane of the present.

This rotation is to be strictly distinguished from a rotation around the spatial axes which includes the time axis. The hypothesis describes the latter as the spin of a particle.

Through the conversion factor c^2, the electric and magnetic field constants can be converted into each other. This also seems to happen when converting the Newton (three dimensions) gravitational constant into the Einstein (four dimensions) gravitational constant [2, p. 202]. There is a gravitational law, which was introduced by Newton, that needed a constant. This law, however, is only valid for a three-dimensional world. Through calculations with differential geometry in the four-dimensional space, Einstein calculated a different gravitational law with its own gravitational constant. However, it is simple to convert these two constants by taking c^4 as a conversion factor. These are all indications of the parallelism and interconnection of the two worlds, which differ only in individual constants by a conversion factor.

THE TIME AXIS

The thin layer of the present is compressed by the gravitation of the time axis. The forces acting on the time axis, exerted by the gravitons, lead to an attraction towards the thin layer, which is the present, from the future as well as from the past. In the layer of the present, there are no forces, since the gravitons, as assumed in this work, have no effect. The masses from which the attractions emanate no -longer lie in front of or behind a particle on the time axis but spatially next to it, thereby exerting no force. The layer of the present will thus become as thin as possible. Additionally, the spatial expansion of the universe is allowed due to the masses distributing without force along the three spatial axes. This could be a possible explanation of dark energy. The term dark energy is generally used to describe the hidden force as the trigger of the expansion of the universe. A thin layer of matter, in which, on an axis perpendicular to the thin layer, its elementary particles attract each other by a force and compress the layer, will flatten out and expand along the other axes upon which no force acts.

Gravity in the sense of mass attraction, such as the attraction of the planets, does not have a direct effect on the body in the form of an interaction but is caused by the curvature of the four-dimensional space-time. This results in an indirect attraction of masses through a change in the three-dimensional space with forces according to Newton's law of gravitation. This curvature of space is caused by the mass of a body. As assumed here, the actual process of the forces takes place on the time axis. This is triggered by the presence of a mass that changes the time axis, but not by the direct exchange of gravitons in three-dimensional space. The force acting on the time axis leads to different physical properties of the time axis depending on the strength of the force. The greater the force, the more compressed the timeline appears physically.

Gravity on the time axis would also be a good explanation for the law of conservation of energy. The basis of the theory of relativity is that all four axes of space-time are equal. As a result, the energy would have to be distributed over all axes, including the time axis;

however, this does not happen. This has already been addressed in the previous chapter. Thus, the energy is only distributed on the three spatial axes.

The changes described by the theory of relativity, which are produced by a velocity or gravitation, are not in contradiction to the hypothesis presented here. A particle's velocity slows down the passage of time in the particle itself by tilting the time axis, as discussed earlier. A force that exists only on the time axis, which emanates from the total mass of our present, creates an effect that is comparable in its physical properties to a contraction of the time axis. The amplitudes of an oscillation on the time axis would vary depending on the size of the potential. This could be compared to a different physical contraction (curvature/tilt) of the time axis.

If the force increases due to the larger mass of a fictional universe, it would correspond to a shortening of the movement displacement on the time axis in the aforementioned example.

Thus, by comparing two otherwise identical master particles (master particles in the same oscillation level), the amplitude of the oscillation of the master particle in a stronger gravitational field would appear more compressed than that of the master particle in a weaker gravitational field. The effect of moving this physical oscillation model at a velocity is the same. The same effect created by a mass can be produced but now by tilting the time axis through a velocity.

The master particle can move both on the spatial axes and on the time axis and can also move faster than the speed of light. It oscillates like a pendulum through a plane that attracts it on both sides. The oscillation remains stable and does not weaken. To put it simply, it is the integer harmonics of the fundamental frequencies that represent the individual energy levels. It seems confusing that the energy belonging to the individual harmonics does not follow integer ratios. However, this is due to the individual overall conditions in which the oscillations are kept stable. A correspondingly complex mathematical formalism is necessary for the calculation.

Since the vibrations caused by the superluminal speed of the master particles create a special feature, we have to add another consideration. The master particles are faster at another point in space than a beam of light. Before any physical interaction with another particle is possible, a kind of proximity to a master particle may well have already been established. An energy transfer that is faster than the speed of light is not possible in our world due to the separation of the two different sections or worlds.

The master particle moves on the time axis and can "try out" all possible ways of interaction and thus perhaps achieve a prioritization. If the movement on the time axis of a master particle results in a certain presence corresponding to the probabilities of its location, an overlap with the probability of a potential reaction partner is also conceivable, given a certain layer of thickness of the presence. According to the laws, a reaction follows with an energy transfer at sub-light speed. One question would be how far along the timeline these influences are possible. At this point, this also gives rise to the question regarding a presumed certain thickness of the present within which the reactions take place.

THE SPEED OF LIGHT

From the basic idea of a one-dimensional force acting on the time axis, the physical behavior of this axis changes to the other three axes of the four-dimensional space-time structure that originally all behaved equally. The speed of light can be expressed as the fixed ratio of the space axes to the time axis and is described as a constant. It directly expresses the ratio of the scales of the axes to one another. Based on this work, the question arises as to whether the force on the time axis determines this ratio and even changes it depending on the magnitude of this force. Does the force on the time axis, and thus also the speed of light, change? And does it depend on the gravitational force emanating from the entire universe? From the general theory of relativity, it follows immediately that the time axis changes as a result of the speed of an object. It is also known that the same effect is produced by a gravitational field. Under these conditions, Einstein's gravitational constant, the mass of the universe and the speed of light should have a mathematically calculable relationship to one another. However, if the speed of light were always constant, the gravitational constant would have to change as a result. It would also be interesting to consider what changes result from the known, constant expansion of the universe. Assuming the speed of light is not a fixed value, a slow constant decrease in the force on the time axis due to the rarefaction of the universe, given by the expansion, would result in a change in the values. Perhaps one day it will be possible to measure this change and answer all these questions.

Suppose that all four axes of space-time have the same unit length in a ratio of 1:1 to each other, without any preference for one axis, but only if no forces are acting on particles that are located on them. Assuming that this scale changes on the time axis in physical behavior due to a curvature or shortening, given a gravitational force, then the relationship of the axes in physical behavior to one another would also change.

The influence of the universe on the timeline should be different depending on the distance from the four-dimensional center of mass. The universe is seen in its entirety as a four-dimensional sphere on the surface of which we are located. Hypothetically, this would mean that the speed of light decreases as the universe expands because the distance to the center of the universe on the time axis increases. Fundamentally, all forces decrease as the distance increases. This holds true for all possible forms of force. This also reduces the gravitational pull of the universe on the present. The center of all mass in the universe is in the past, and the distance from the center of mass on the time axis determines the strength of gravity. See Appendix 2 for a more detailed explanation and furthering of this idea.

ASSIGNING DIFFERENT ENERGY LEVELS TO FERMIONS

The individual stable energy states in which the master particles occur each represent different fermions with the common properties of the whole group. This is the case, for example, with all leptons (electron, muon and tauon). However, the ground state of the associated master particle, the one with the lowest energy or the first energy state in the parallel world, can be assigned to a boson. As we will see later, it might be composed of two particles: a master particle and its antiparticle. This leads to the possibility of a) justifying the mass of the bosons Z and W, and b) explaining the transition of the fermions into each other by the exchange of a W boson. This is shown in the next section for the radioactive beta decay of bound fermions (e.g., the quarks in a neutron or proton).

Interestingly, there are conservation laws for the number of leptons and the same for the number of quarks. The sum of all leptons minus the antileptons (including the neutrinos and antineutrinos) remains the same after a particle reaction and is thus equal to the total number before the reaction. This also applies to the sum of the two quark groups and their antiparticles. The idea that for each group there must only be one master particle or anti-master-particle in different oscillation levels arose due to that fact, among others.

Back to the oscillation energies in the individual stable oscillation levels. The basic oscillation in the parallel world for the generations of leptons corresponds to one boson, i.e., the lowest energy level of the generations. The next energy level up is assigned to the heaviest fermion of the group, the tauon (see Fig. 1). This continues with the muon, and finally, the highest energy level of the oscillation in the parallel world is assigned to the electron. The stability of the fourth and last energy level of the electron cannot be easily explained. There could exist further energy levels that are close to it. However, this may not be the case because of other properties of the real world, such as the charge or the spin of the electron itself, which contain energy that cannot be surpassed.

Increasing the speed of a particle, both in the real world and in the parallel world, leads to an increase in energy. This happens from the opposite direction. Speed in the parallel world is defined inversely (e.g., s/m), and changing the speed results in an increase or decrease in energy. This creates an opposing development of the energy increases of both worlds. The more energy a master particle has in the parallel world, the lower the mass of a fermion in the real world. This statement is the result of many attempts to interpret the calculation results with the formula used in this book. What is astonishing is that as a result, the energy levels with the highest energy in the parallel world are the desired and most stable states, unlike in the real world. In our world, the particle with the lowest energy or mass is the most stable. However, from this, it follows that, e.g., the W boson, which is the heaviest particle in our world, has the least amount of energy in the parallel world. Conversely, in the parallel world, the lightest particle in the group, the electron, consists of the master particle with the greatest energy bound in the oscillation. This seemingly insurmountable contradiction can provide an explanation for the existing but incomprehensible energy if it is assumed that this difference is responsible for dark matter. This becomes clear when, together, the lightest fermions determine almost all of the visible mass of the universe, which in turn only corresponds to a small part of the total mass of the universe. The larger part is attributed to dark matter. However, this could correspond to the mass that aligns with the bound energy of the master particles in the parallel world. Additionally, we should also consider that there is no absolute nothingness in a vacuum. Due to pair production and annihilation, there is an abundance of particles, with dark matter occurring most frequently, especially in the seeming "nothingness". The lighter elementary particles are predominantly produced here; however, they form the greatest energy in the parallel space and thus most of the dark matter. For example, the most common fermions with the highest energy in the parallel world are electrons. This is a newly developed hypothesis from an overall consideration of the discussed observations and their connections.

There would be a hypothetical explanation for the stability of the highest energy state in the parallel world. It can be understood by imagining that the attractive interaction of gravity on the energy of a particle creates a tendency to maximize the energy in the parallel world. This is quite plausible, as energy and mass are seen as equivalent, attract each other and thus tend to increase in all cases. These would be the stable states in the parallel world, as opposed to the inverse ratios in the real world. In other words, there will always be dark matter maximization in all particle processes.

In the parallel world, energy is attracted by the force on the time axis and is thus also held in the highest possible energy level. This explains the decay of the heavier fermions into the lighter ones in a different way. This increases the dark matter, as it is lower for the heavy fermions. They exist in a lower energy state in the parallel world. On the other hand, it means that fermions that absorb energy in the real world would equate to less dark matter in the parallel world. There are a number of questions arising from this assumption that would be interesting to find the answer to. For example, what happens to the dark matter when a particle and an antiparticle annihilate? This energy would still have to follow the law of energy conservation. How can it be observed or assumed?

RESULTS AND IDEAS TO CALCULATE THE MASS OF A FERMION

The most difficult part will be to find calculations for the different oscillation energies. A quantum mechanical solution is a requirement and prerequisite for this, and it will be made possible with different mathematical approaches. An idea would be to use the calculation model of a falling object. That would be simple and could be used as a first approximation of the mass calculations. There are definitely better ideas for this kind of calculation, and they will have to follow. In the following approach in this work, the previously discussed intricacies, such as those of angular momenta, have not yet been accommodated. Many difficult and, above all, complex new mathematical ways have to be found in order to describe the idea in a physically and mathematically correct manner.

The idea of a master particle moving away from and coming back to the three-dimensional plane while moving on the time axis is compared to a ball falling to earth from a certain height. In order to get an overview of the results to be expected, the solution of a Schrödinger equation is chosen for the quantum mechanical solution of the problem. In the Schrödinger equation, the state of the system is represented by a wave function [15, p. 95]. There is a mathematically complex way to do this, which is described in Task 33 in [16, p. 92—95].

$$-\frac{\hbar}{2m}\frac{d^2u}{dx^2} + m \cdot g \cdot x \cdot u = E \cdot u$$

Here, $m \cdot g \cdot x$ is the potential of gravity, u is a required solution function and the boundary conditions are (0) = 0 and (∞) = 0.

The quantum mechanical calculation of the free fall, taken as a basis, may lead to a first estimation when applied to the different masses of the fermions. The three different master particles lead to three different oscillation models: one for the electron, muon, tauon and *W* boson; one for the three quarks d, s, b and, belonging to the lowest oscillation level, the Higgs boson; then one more for the quarks u, c and t, with another yet unknown, and assumed very heavy, boson. The proof of the existence of such a boson would cement the evidence of this hypothesis. The task of discovering the new boson is quite a difficult one, as the boson would have to have a very large mass of around 600—900 tera electron Volts, which can be assumed from the calculations.

It is possible that there are only two master particles: one for the leptons and one for the two quark groups. This conclusion arises from the conservation of quark number, which remains constant in particle reactions for both groups.

Overall, taking into account any discrepancies, the results according to the formula lead to a mass distribution (see tables in Appendix 1) that corresponds to real observations.

This view of the connections, however, leads to a completely new position of the Higgs particle. So far, it has been seen as a superordinate particle. This cannot be solely inferred from its position here, as its mass aligns precisely with the previously mentioned position of the boson in the series of quarks d, s and b. It would be comparable in significance to the *W*± or *Z* particle. The main decay of the Higgs particle into quark b and the antiquark of b corresponds to the transition to the neighboring position in the same group. Another decay path to the other bosons results from the initial energy level of the Higgs boson in the energy level sequence of its own group. The Higgs boson at the beginning of the group of quarks is transformed into a *W* boson, which is also at the beginning of a (different) group. The beginning of the group of leptons is determined by the *W* particle, which in principle can be reached by other particles, as the position at the beginning of the group is essential for this. From this new viewpoint, other connections arise and give the particles a new meaning. The fundamental oscillations are the bosons (*W* and *Z*, the Higgs boson and an unknown new boson), and the new and unknown boson could have a superordinate position like the Higgs boson. The difference between a fermion and a boson, hypothetically according to this theory, could lie in the possibility that bosons contain a master particle and an anti-master-particle at the same time, bound in the same oscillation level. This would be the easiest way to explain the decay of the W boson or Z boson into a lepton and an antilepton (antineutrino). The decay of the Higgs boson into a quark b and an antiquark b would also fit well with this assumption. The master particles and anti-master-particles could rotate in the same direction, counterclockwise or clockwise (charge 1 or minus 1), or they could rotate in the opposite direction (zero charge) around the time axis. This way the charge of 1, minus 1 and 0 (opposite rotational direction of both particles) could be explained. The same considerations can also be made analogously for the spin of the master particle. In the following chapter, Figure 2 illustrates this by using radioactive beta decay.

RADIOACTIVE BETA DECAY AND HYPOTHESIS

In a beta decay, a nucleus transforms into a nucleus of another chemical element. In a β^- decay, this is the neighbor element with a higher atomic number, and in a β^+ decay, it will be the neighbor element with a lower atomic number. Beta radiation is particle radiation. In the case of β^- radiation, it consists of negatively charged electrons, and in the case of β^+ radiation, it consists of positively charged positrons. In addition, an antineutrino is released in a β^- decay and a neutrino in a β^+ decay.

The name comes from the very first classification of ionizing rays from radioactive decay into alpha, beta and gamma rays, which show increasingly penetrating power in that order.

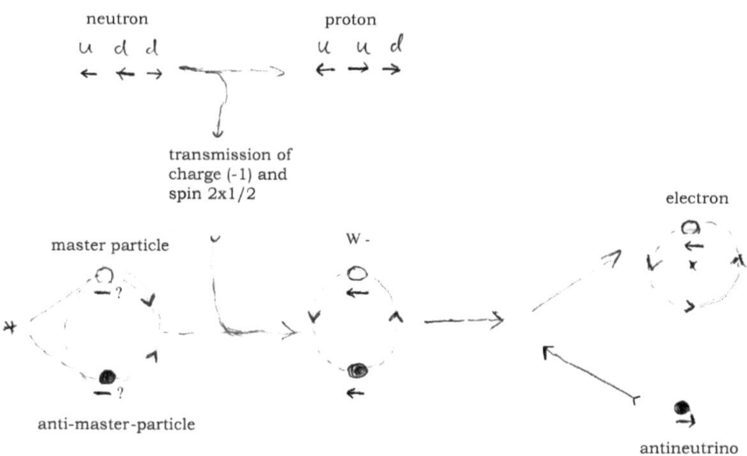

Figure 2: The beta decay, the small arrows under the particles indicate the spin direction, the dashed circles with the direction of rotation indicate the direction of rotation around the time axis (counterclockwise for negative electric charge)

Due to the idea behind this work, the beta decay could follow the diagram in Figure 2:

A master particle and an anti-master-particle result from a spontaneous pair production. Both rotate in opposite directions around the time axis. Overall, this corresponds to a neutral charge. The spin or intrinsic angular momentum is very important in the further course of the reaction and will always be the same for reasons that are still unclear, with a right-handed spin for the anti-master-particle and a left-handed spin for the master particle. However, it is possible that this is not yet determined at this point but only becomes necessary later in the course of the reaction. In this configuration, the spins of both particles, as well as the rotations of the two particles around the time axis, cancel out. How the further transitions to a conversion into a neutrino and antineutrino take place remains open. But master particles and neutrinos could be identical particles. Continuing from the process outlined above, this particle and anti-master-particle pair then receives a negative charge and an integer spin through a left-handed (spin) reaction partner, which is a d quark (charge -1/3, spin -½), by the conversion to a u quark (charge 2/3, spin ½). The quarks are components in a nearby neutron of an atomic nucleus. This decays through beta decay. Referring back to Figure 2, the charge causes the particle and the antiparticle (depending on the negative or positive charge, left or right) to now rotate in the same direction around the time axis. The negative charge means that the particles now rotate counterclockwise around the time axis. Thus, a W^- boson is created. The master particle and its antiparticle are treated as a single particle with respect to charge. It would also be possible that the charge is only created by one of the particles, while the other is not involved. A close examination would reveal the properties of the charge of the W boson and whether there is anything special about this charge. This boson only exists for a short time and decays into a right-handed antineutrino and a left-handed (this results from the direction of motion of the electron and is also experimentally confirmed) electron. In this process, the W boson would thus be a vehicle for charges and other physical properties such as angular momentums. In addition to a possible explanation of beta decay, the lepton conservation law is also a fact that is self-explanatory through this process. At the same time, the number of quarks and antiquarks remained the same and the compliance with the conservation law is also given.

The interaction with the W^- boson is called the weak interaction. It is a separate force from the other three interactions, such as gravity. In beta decay, for example, it transforms a down quark into an up quark. The transfer of charges in beta decay is only possible through weak interactions, whereby the charge conservation law is complied with. The weak interaction is not mirror symmetric. Certain processes of the weak force only take place with particles with a certain spin orientation. In the case of fermions, the spin, like the mass, is an invariable inner particle property.

In beta decay, only one left-handed quark reactant is used, and the result is always a left-handed electron and right-handed antineutrino, as shown in Figure 2 [8, p. 214—224]. The mirrored process with opposite spin orientation does not occur in beta decay. This phenomenon is known as parity violation, as physical processes are otherwise independent of spatial preferences. The weak interaction also differs between particles and antiparticles, as the preference for a spin direction for particles is the exact opposite for antiparticles. The spin is a half or integer multiple of the reduced Planck constant $\hbar$, and the spin has all the properties of a classic mechanical intrinsic angular momentum. The law of conservation of angular momentum also applies to the spin for all involved particle reactions.

The spin of a neutrino is always left-handed, while for antineutrinos it is always right-handed. So far, this is inexplicable and quite special in particle physics.

What may play a role in beta decay is that a particle's spin must match the particle's charge in a non-interfering manner. According to the hypothesis in this book, the charge and spin of a particle are both angular momentums of the particle but around different axes, which could well interfere with each other.

SUMMARY

In this book, an explanation for the cause of the mass of elementary particles and their antiparticles as well as their mass ratios to each other is sought. A new Standard Model is developed. A Standard Model [1] is a summary of all the properties of the known elementary particles and the most important forces between them.

It is assumed and substantiated that the elementary particles, i.e., fermions and bosons, consist of an even smaller particle. This particle is introduced as a master particle. The only difference between different fermions is the different energy levels a master particle can be in, which then creates the different known fermions. This creates a new Standard Model of elementary particles. It is justified under the assumption that the master particle oscillates on the time axis. This is a segment that exists beyond the speed of light and is thus inaccessible from our real space. With that, the prerequisites from the basics of the theory of relativity are met. Something that is moving at superluminal speed would not be observable to us and thus would not interact with us. It is, so to say, not feasible. The master particle, under this hypothetical assumption, moves at a speed exceeding the speed of light on the time axis. Mathematically speaking, this assumption of a particle surpassing the speed of light would not contradict the theory of relativity. The physical processes are mathematically transferrable to the superluminal segment. A parallel-existing reality would be possible without it having any references to the world we perceive as real.

The gravitons play a special role. In the introduced hypothesis, the gravitons only act on the time axis and are responsible for the force of creating the motion of master particles. The gravitons cannot exert any forces on one or any of the spatial axes, which would then explain why they haven't been observed yet, as they can only be found on the time axis. Including gravitons in the Standard Model is a novelty and has so far not been possible.

Due to the hypothesis, a new understanding of the connections between different quantum mechanical phenomena is possible.

Quantum mechanics is a projection of the four-dimensional reality onto the three-dimensional world known to us. A hypothetical explanation for Heisenberg's uncertainty principle would be an angular momentum around the spatial axes of a master particle oscillating on the time axis. This means that a master particle can only go below a certain radius around this center in the three-dimensional plane through high energies or for a very short time. There are no limits to larger radii, only a decreasing probability of presence with an increasing radius.

The tunnel effect can also be explained due to the movement of a master particle on the time axis. The master particle would be able to overcome a potential barrier or obstacle through a detour on the time axis. The particle could go from one place to the next across the time axis without having to go over the potential barrier existing on the space axis. A different rotation of the master particle, this time around the time axis, is possible as a more profound explanation for an electric charge. With a projection of this rotation onto the real world, the resulting phenomena are consistent with the Maxwell equations. The rotation of the particle around the time axis would correspond to the rotation of the magnetic field around the direction of movement of the particle. This would very closely correspond to the phenomena of magnetism described by Maxwell. This also leads to an explanation of a missing magnetic monopole, as the rotation is self-contained. A magnetic monopole would be a sole magnetic north pole with no south pole, or vice versa. There is also no actual magnetic charge. A magnetic field is always self-contained with a north and south pole. The magnetic field is perpendicular to the direction of motion and rotates around the moving charge.

A well-known simple conversion of physical constants of the three-dimensional and four-dimensional worlds into one another suggests that there are also other connections between the two worlds. This is evident, for example, when taking the bisquare of the speed of light (c^4) as a conversion factor for the two known gravitational constants in each space. The consequence of all the connections described is that two worlds exist, separated into one for phenomena slower than the speed of light and one for phenomena caused by movements faster than the speed of light. These two worlds cannot be considered independently. There is a link between both worlds through and within the elementary particles with mass.

The energy in the parallel world is the indirect cause of the mass of the elementary particles. It is bound in the oscillation of the master particle on the time axis. The more energy a master particle has in this parallel world, the lower the mass of the fermion in the real world.

This energy interacts with other energies and the associated force is mediated through gravity. Unlike in our three-dimensional world, in the parallel world, the energy will always try to reach a maximum. Here, in our world, the lowest energy state will always be the most stable. An unequal ratio of energies between the two worlds results in a difference that leads to an energy surplus in the parallel world, which could provide a possible explanation for dark matter.

The status of bosons in the Standard Model would also be changing. Particles that mediate forces are called bosons and are assumed to be composed of a particle and its antiparticle. They are seen as carriers of physical properties. The difference between a fermion and a boson might lie in the possibility that bosons contain a master particle and an anti-master-particle at the same time, bound in the same oscillation level. In contrast, fermions consist of only one master particle.

The possible detection of another unknown boson may one day be particularly interesting. It could fit into the developed model and be a validation of the hypothesis. The existence is suspected based on this work.

A conversion of fermions within a group with a generation change would be simple. This would be the result of adding or subtracting energy. However, the transition of one quark from one quark group to the next is harder to explain. However, this is possible due to the weak interaction with the exchange of a W boson, as described previously. The weak interaction is the only force that can transform quarks from different groups into each other. This interaction is therefore particularly interesting for this work, as an explanation for this becomes possible with the introduced idea. The crucial principle is that the bosons can convert elementary particles into each other. Further transitions with the other bosons are to be expected according to the processes developed.

There are a total of four known bosons with mass, W^+, W^-, Z and Higgs, and perhaps another yet to be discovered, which is postulated in this book.

Due to a compression on the time axis, the universe expands three-dimensionally. This is due to the one-dimensional force of gravity acting on the time axis. This is the reasoning behind this work.

Many questions need to be addressed, including whether a possible change in the physical constants, in particular the speed of light, will be observed with the increasing age of the universe. Due to the ideas within this book, it is to be assumed.

The acceptance of the working hypothesis is not as simple; however, it is plausible to put the previously unexplained phenomena in physics together in a different way.

APPENDIX 1

In order to get a rough overview, the solution of a Schrödinger equation is chosen for the calculation in the classical sense. In principle, this approach is not suitable for accurate results when dealing with particles with a spin. However, the spin probably does not have much influence on the calculation of the mass of the fermions, and so this way is chosen for a rough approximation.

This approach has already been addressed and implemented using the following formula [16, p. 92—95]:

$$-\frac{\hbar}{2m}\frac{d^2u}{dx^2} + m \cdot g \cdot x \cdot u = E \cdot u$$

From this, the formula for the energy is calculated using a length, l, the mass, m, of the master particle and a dimensionless parameter, λ:

$$E_\lambda = \frac{\hbar^2}{2ml^2}\lambda$$

And with the following eigenvalues for the energy, λ_1 = 2.33; λ_2 = 4.08; λ_3 = 5.51; λ_4 = 6.78.

By converting the potential energy of a master particle into kinetic energy, the maximum amount of kinetic energy is reached when passing through the three-dimensional hyperplane of the present. The individual parameters (l, m and λ) were determined by inference using the known mass of the fermions.

Table 1 contains the values of the masses for the leptons: electron, muon and tauon, and the energy levels with the eigenvalues λ. This table has only been created to illustrate the intended relationships without a mathematical background.

For the sake of completeness, corresponding tables of the parameters can be found in Tables 3 and 4 for the two groups of quarks.

Only for the position of the unknown boson, an estimated calculation is derived from the parameters of the group of quarks in Table 4.

In the following tables, the correspondences between the eigenvalues λ of the solution function of the aforementioned Schrödinger equation and the energy values (mass) of the individual fermions are listed.

Table 2 shows the values for the electron and the W boson calculated with the Schrödinger equation if the initial conditions were determined by the known masses of the muon and tauon. It is to be noted that the calculated mass for the W boson was only half as large as expected. This supports the assumption that this boson consists of two particles, as already indicated in the text.

A calculation of the energies in the parallel world (dark matter) was pursued but not listed here. This is due to too many speculative assumptions that have to be made during the calculation process.

Table 1: Summary of the known parameters and the eigenvalues λ of the solution function of the Schrödinger equation.

Fermion (lepton)	**Electron**	**Muon**	**Tauon**	*W* **boson**
Mass (MeV)	0.511	105.7	1777	80400
λ	6.78	5.51	4.08	2.32

Table 2: The parameters correspond to those in Table 1. In Table 2, the masses of the electron and *W* boson have been calculated using the formula given previously based on the masses of the muons and tauons.

Fermion (lepton)	**Electron**	**Muon**	**Tauon**	*W* **boson**
Mass (MeV)	8	105.7	1777	48000
λ	6.78	5.51	4.08	2.32

Table 3: Summary of the known parameters for the group of the three lighter quarks.

Fermion (quark)	**d**	*s*	**b**	**Higgs boson**
Mass (MeV)	4.8 (5-8.5)	104 (80-155)	4700 (4000-4500)	125000
λ	6.78	5.51	4.08	2.32

Table 4: Summary of the parameters known for the second group of quarks and the calculated parameter for the mass of an unknown, new boson.

Fermion (quark)	**u**	**c**	**t**	**(?) boson**
Mass (MeV)	2.4 (1.5-8.5)	1270 (1000-1400)	171000	600-900 $\cdot$ 10^6
λ	6.78	5.51	4.08	2.32

APPENDIX 2

$$\frac{1\ \textit{unit of length on the time axis}}{1 + \eta\ \textit{force on the time axis}}$$

Given a unit of length and the force on the time axis that is acting on a particle, a resulting change in the physical property of the time axis could be described as shown in the formula above. Here η is a constant of proportionality.

This is calculated correspondingly for the spatial axes given the following formula:

$$\frac{1\ \textit{unit of length on the spatial axis}}{1 + \eta\ \textit{force on the spatial axis}}$$

The ratio of both quotients to each other could determine the speed of light if all space axes and the time axis are originally to be considered physically equivalent:

$$\frac{1\ \textit{unit of length on the spatial axis}}{1\ \textit{unit of length on the time axis}} = \frac{1 + \eta\ \textit{force on the time axis}}{1 + \eta\ \textit{force on the spatial axis}}$$

Assuming that there is no acceleration (= 0) on any of the space axes and the time axis, the relation between the space and time axes results in a ratio of 1. This relation would arise if the space axis, e.g., x, and time axis t were equal.

A gravitational force acting only on the time axis would physically contract the unit length of the time axis but not the x axis. This change is dependent on the gravitational force emanating from the mass of the entire universe.

$$\frac{1\ \textit{unit of length on the spatial axis}}{1\ \textit{unit of length on the time axis}} = c$$

$$c = \frac{1 + \eta\ \textit{force on the time axis}}{1}$$

c is the speed of light.

   Assuming the force on the spatial axis = 0 and = 1, with the value to be estimated for the force on the time axis (with the dimension of acceleration on the time axis), the law of gravitation yields:

$$force\ on\ the\ time\ axis\ =\ \kappa \cdot \frac{mass\ of\ the\ universe}{s^2}$$

In this formula, s = 1 (a four-dimensional line element), the mass of the universe = $1.45 \cdot 10^{51}$ kg and Einstein's gravitational constant

$$\kappa = 8 \cdot \pi \cdot \frac{1}{c^4} \cdot 6{,}67 \cdot 10^{-11} \{m^3 \cdot \frac{1}{kgs^2}\}.$$

Substituting the values of the parameters yields:

$$c = 1 + 1.45 \cdot 10^{51} \cdot 6.67 \cdot 10^{-11} \cdot 8\pi \frac{1}{c^4} = 2.9 \cdot 10^8 \ \{\frac{m}{s}\}$$

The magnitude given for the mass of the universe is within the range of known assumptions. With this rough calculation, the speed of light can be estimated using the gravitational constant and the mass of the universe.

BIBLIOGRAPHY

[1] M. Thomson, Modern Particle Physics, 1st ed., New York Cambridge: Cambridge University Press, 2013.

[2] W. Pauli, Relativitätstheorie, 1st ed., Torino: Paolo Boringhieri, 1963.

[3] W. Rindler, Relativity, 2nd ed., Oxford New York: Oxford University Press, 2008.

[4] R. Shankar, Principles of Quantum Mechanics, 2nd ed., New Delhi: Springer Verlag, 1980.

[5] A. Messiah, Quantenmechanik 1, 2nd ed., vol. 1, Berlin New York: Walter de Gruyter, 1991.

[6] A. Messiah, Quantenmechanik 2, 3rd ed., vol. 2, Berlin New York: Walter de Gruyter, 1990.

[7] W. Nolting, Grundkurs Theoretische Physik 5/2, 6th ed., vol. 5/2, Berlin Heidelberg: Springer Verlag, 2006.

[8] K. W. Ford, Die Welt der Elementarteilchen, 1st ed., Berlin Heidelberg: Springer-Verlag, 1966.

[9] W. Nolting, Grundkurs Theoretische Physik 3, 8th ed., vol. 3, Berlin Heidelberg: Springer Verlag, 2007.

[10] R. H. Sanders, The Dark Matter Problem, 1st ed., New York Cambridge: Cambridge University Press, 2010.

[11] W. Finkelnburg, Einführung in die Atomphysik, 12th ed., Berlin Heidelberg: Springer Verlag, 1976.

[12] C. W. Misner, K. S. Thorne and J. A. Wheeler, Gravitation, 3rd ed., New York: W.H. Freeman and Company, 1973.

[13] C. Kiefer, Quantum Gravity, 2nd ed., Oxford: Oxford Science Publication, 2007.

[14] P. Raschewski, Riemannsche Geometrie und Tensoranalysis, 2nd ed., Frankfurt am Main: Harri Deutsch, 1995.

[15] W. Nolting, Grundkurs Theoretische Physik 5/1, 6th ed., vol. 5/1, Berlin Heidelberg: Springer Verlag, 2004.

[16] S. Flügge, Rechenmethoden der Quantentheorie, 6th ed., vol. 33, Berlin Heidelberg: Springer Verlag, 1999.